MW00908359

ENDANGERED!

MANATEES & DUGONGS

Amanda Harman

Series Consultant: James G. Doherty
General Curator, The Bronx Zoo, New York

BENCHMARK BOOKS

MARSHALL CAVENDISH
NEW YORK

Benchmark Books
Marshall Cavendish Corporation
99 White Plains Road
Tarrytown, New York 10591-9001

Library of Congress Cataloging-in-Publication Data

Harman, Amanda, 1968-
Manatees & Dugongs / by Amanda Harman.
p. cm. — (Endangered!)
Includes index.
Summary: Describes the physical characteristics, habitat, and behavior of manatees and dugongs—endangered, vegetarian mammals that live in the water.
ISBN 0-7614-0294-2
1. Manatees—Juvenile literature. 2. Dugong—Juvenile literature. 3. Endangered species—Juvenile literature. [1. Manatees. 2. Dugong. 3. Endangered species.] I. Title. II. Series.
QL737.S6H37 1997
599.5'5—dc20
96-7220
CIP
AC

Printed and bound in the United States

PICTURE CREDITS

The publishers would like to thank the following picture libraries for supplying the photographs used in this book: Ardea 13, 17, 18, 23; Bruce Coleman Ltd 1, 4, 7, 10, 19, 22, 24, 25, 29; Frank Lane Picture Agency 5, 20, 26, 28, BC; Frank Lane Picture Agency/Silvestris 8, 15; Natural History Photographic Agency FC, 11, 12, 21; Oxford Scientific Films 6, 14, 27.

Series created by Brown Packaging

Front cover: West Indian manatee.
Title page: West Indian manatee.
Back cover: Dugong.

Contents

Introduction

A West Indian manatee swims in a Florida river. Manatees usually have smaller heads than dugongs do, with a shorter snout.

If you met a manatee or a dugong gliding along in its underwater home, you might think it looked like a cross between a whale and a seal. These big, peaceful creatures have a plump, rounded body, two front flippers, and a large flat tail. Their face is gentle-looking with a long, blunt, wrinkled snout, tiny eyes, crescent-shaped nostrils, and no visible ears. The snout is fringed with bristly hairs.

Although they live all their life in water, never venturing out onto dry land, manatees and dugongs are not fish but belong to a large group of animals called **mammals**. Whales and dolphins also belong to this group, as do dogs, cats, human beings, elephants, and many others. In fact, scientists believe that elephants may be the closest living relatives the manatees and dugongs have. Because they are mammals, manatees and dugongs cannot breathe underwater, but must come to the surface to gulp fresh air.

Manatees and dugongs look very similar, but there are some differences between them. For a start, manatees are usually a little larger than dugongs. Both male and female manatees may grow up to 15 feet (4.5 m) long and may weigh as much as 3500 pounds (1600 kg). Dugongs, on the

A manatee comes to the surface to take a breath. It closes its nostrils when it dives so that it does not get water in its lungs.

other hand, measure up to 13 feet (4 m) in length and weigh up to 2000 pounds (900 kg). Also, manatees' tails are very rounded, whereas the tail of a dugong is shaped more like that of a whale. Unlike manatees, dugongs have two long teeth that stick out of their mouths like small tusks. And some manatees have small "fingernails" on the end of their flippers, which dugongs do not.

The dugong and the three **species** of manatees belong to a group of mammals called sirenians (sy-REE-nee-uhns). Not so long ago, there was another creature that belonged to this group. This was Steller's sea cow, a massive creature that was three times as long and weighed six times

A pair of West Indian manatees greet each other. You can see the fingernails on the ends of their flippers.

as much as the dugong. It was first discovered in 1741 in the North Pacific between Alaska and Siberia. Right away, whale hunters and seal hunters began killing this animal for its meat. They also took the blubber, the layer of fat beneath the animals' skin that protected them from the cold. This contained oil, which was used for a number of purposes, including lighting. Within just 30 years of being discovered, Steller's sea cow was completely **extinct**.

A small group of fish swim ahead of a dugong. The fish are in no danger from the dugong since it eats nothing but plants.

Sadly, it seems that manatees and dugongs may be in danger of vanishing from the planet, too. In this book we will discover how these shy, harmless creatures live, and why their survival is at risk. Let us begin with dugongs.

Dugongs

The dugong is found along the coasts of east Africa, Asia, New Guinea, and Australia. It also used to live off northern Madagascar, although it is probably extinct there now. In spite of its size and bulk, the dugong's body is well **adapted** to its watery **habitat**. It is widest in the middle and tapers toward the head and tail, and the skin is smooth and hairless. As a result, the dugong can move through the water with ease. Most of the time, it swims slowly and lazily. But the dugong can move quickly when it needs to.

A dugong in clear, sunlit waters. Dugongs swim by moving their powerful tail up and down and steering with their flippers.

Beneath its tough skin, the dugong has a thick layer of blubber, which helps to keep it warm. Even so, the dugong needs warm water to survive and lives only in **tropical** regions. It is not a deep diver and can usually be found close to the shore in places where the water is about 30 feet (9 m) deep. The dugong is hardly ever seen swimming out in the open ocean.

Dugongs eat at all times of the day and night. They are **herbivores**, which means that they do not eat meat, but feed on plants. In fact, dugongs (and their close relatives the manatees) are the only sea mammals that are completely **vegetarian**. However, because of their large

Areas where the dugong can be found

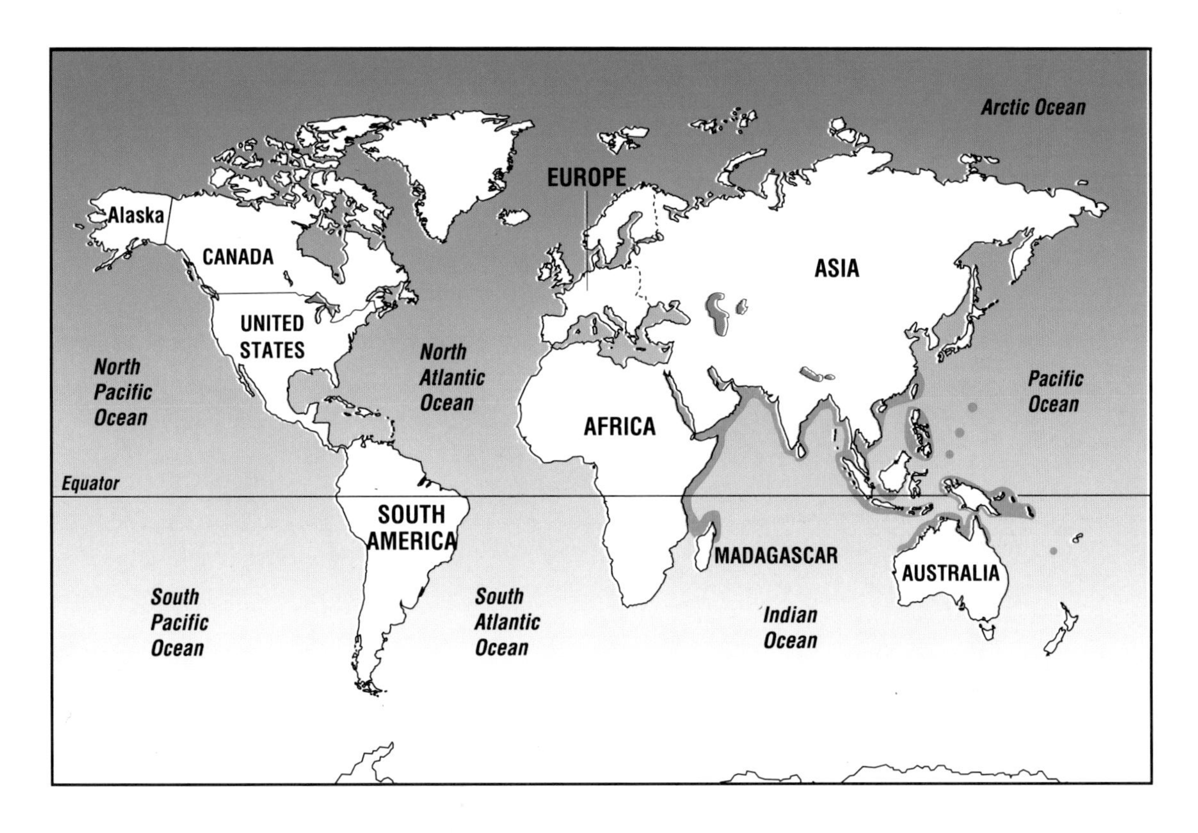

size, these animals have to eat a huge amount of food to keep alive. They probably eat about 70 pounds (32 kg) of water plants every day.

The dugong has the nickname "sea pig" because it likes to root around for food on the seabed with its big, bristly snout. Its favorite food is a type of plant called seagrass, especially the juicy roots, which it digs up using its flippers. Besides their "tusks," which do not seem to be used for feeding, dugongs have just a few teeth in the back of their jaws. These are like little pegs, and since they are not of much use, dugongs rely mostly on two thick horny plates lining the top and bottom of their mouth to chew

A dugong comes to the surface during a meal of seagrass. Dugongs need to come up for air about once a minute while they are feeding.

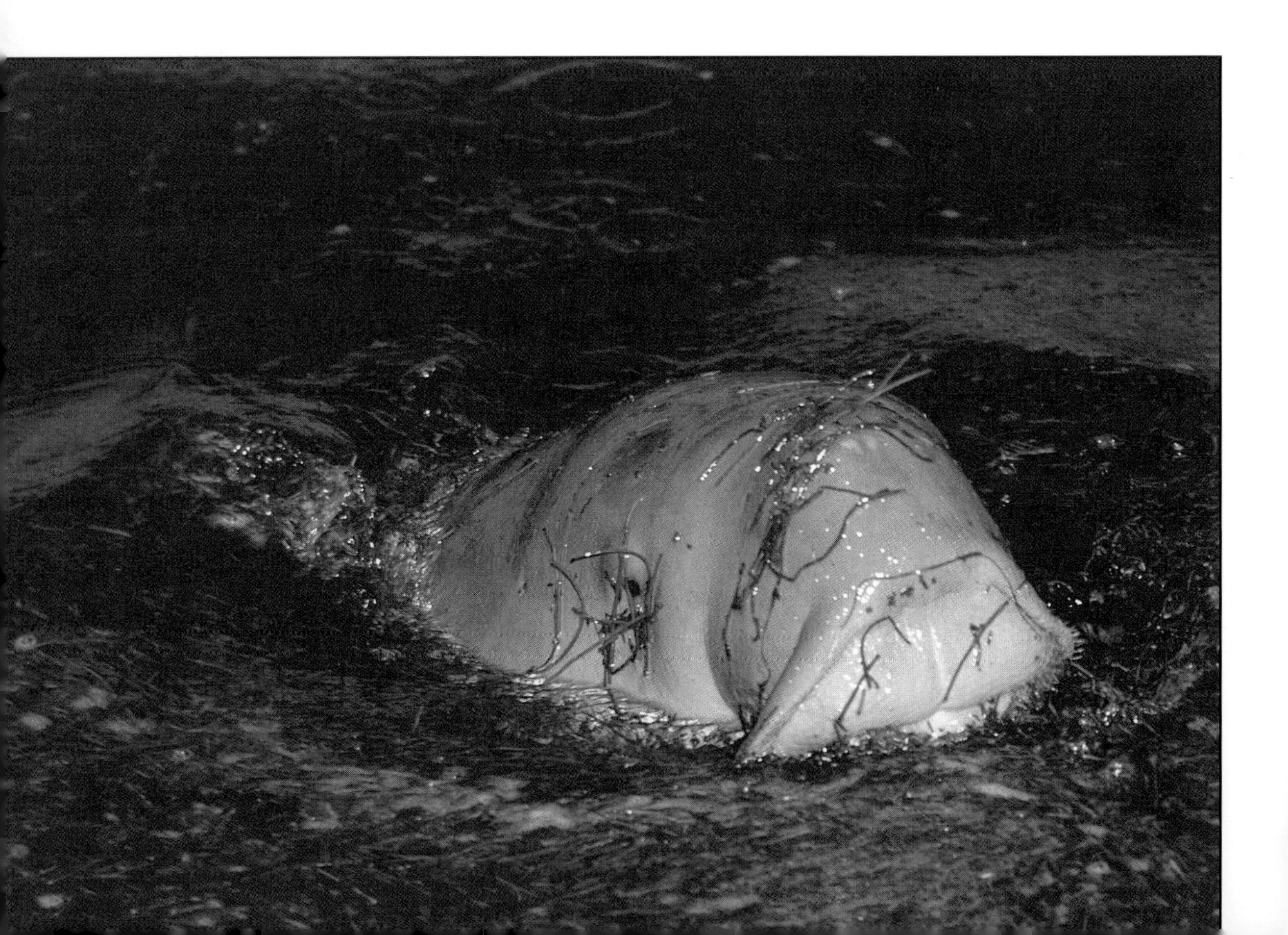

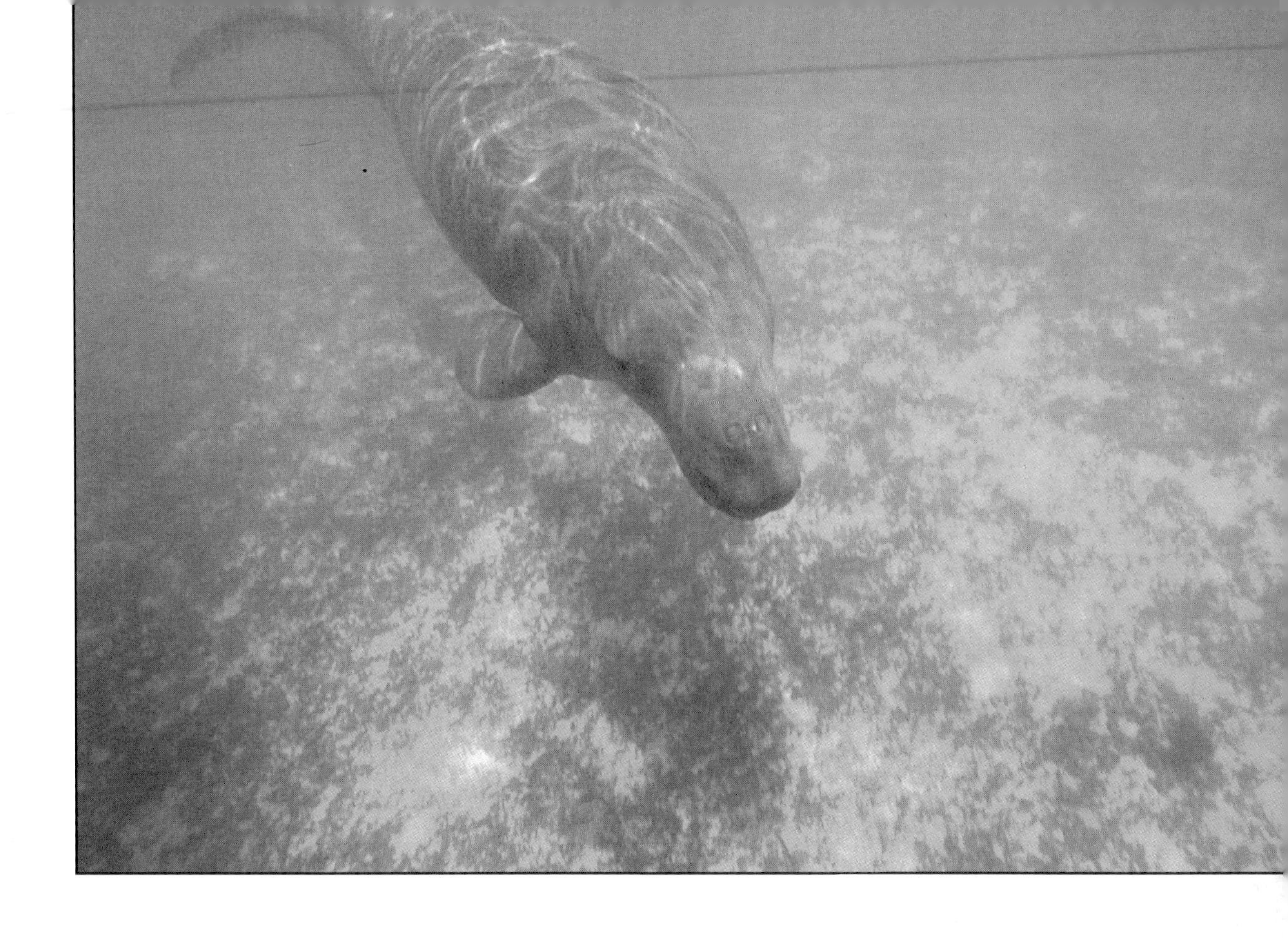

their food. Very few animals compete for food with dugongs. In fact, the only other large creatures that feed on seagrass are sea turtles.

At one time, dugongs were regularly seen, often in huge numbers. There are reports of dugong herds off the coast of Australia that spread more than 3 miles (5 km) long and 900 feet (275 m) across! Then people started to hunt dugongs on a large scale, and today they are hardly ever seen at all. This is partly because dugongs tend to spend most of their time alone now, rather than in large groups. The only time dugongs get together in herds these days is when the temperature drops in winter and the water

A dugong cruises over a seagrass bed in the warm waters of the South Pacific. Dugongs travel up to 16 miles (25 km) every 24 hours as they feed.

Dugongs are sometimes attacked by sharks, and this one has had part of its tail bitten off. However, since they are so big, dugongs are not easy to kill. They have been known to butt sharks with their heads to drive them away.

becomes very cold. Then up to a hundred animals will gather where the water is warmer. Also dugongs have become very shy. Sometimes they will still come to check out a diver or a boat that enters their area. But usually dugongs keep away from humans as much as possible and come to the surface of the sea only to breathe.

The dugong breeding season begins in September, when each adult male claims an area of sea called a **territory**. Males defend this area against rivals. Female dugongs swim through the territories, looking for a suitable partner. Females are usually able to **mate** by the time they are about nine or ten years old, although some are not ready until they reach 18. As the females swim by, the males try to impress them with how fit they are by performing "sit-ups" in the water. The male thrusts himself up out of the water energetically and then falls back with a splash.

Once this display is over a female mates with her chosen partner or partners. About a year later she gives birth in the water to a single calf. The baby dugong is usually about 35 inches (89 cm) long and weighs about 40 pounds (18 kg). As soon as it is born, its mother swims beneath it so that it rests on her back. Then she swims quickly up to the surface, where the baby takes its first tiny gulp of air.

A dugong in calm waters off Australia. Dugongs avoid rough seas since they may be washed ashore and then find it difficult to get back into the water.

Dugong calves stay with their mothers for up to two years. Mother and young have a close relationship during this time, and the two are almost always together. The mother teaches her offspring all it needs to know, such as how to feed and where to find the best food. Young dugongs are able to feed on seagrass within a few weeks of being born, but they continue to drink their mother's milk until they are almost old enough to start lives of their own. There are stories of dugong mothers cradling their babies in their flippers while they **suckle**, but these are probably not true. The mother's teats are located just behind her flippers, and while it feeds, the baby usually swims along by her underside with its belly facing upward.

Dugongs, like manatees, are not very noisy animals, but they do sometimes call to one another with squeaks and grunts.

Manatees

The three species of manatees are the West Indian, the West African, and the Amazonian. The West Indian is found throughout the Caribbean Sea. Its **range** also reaches as far north as the southeastern United States and as far south as northern South America. The West African manatee is found along the west coast of Africa from Senegal to Angola. Both of these species are gray-brown in color and have "fingernails" on the tips of their flippers. Thousands of years ago, when manatees lived on land, these nails were probably claws or hooves. Over time, the animals used them less and less until they withered away.

West Indian manatees have been seen in Chesapeake Bay. Usually, though, they travel no farther north than Florida.

Unlike their cousin the dugong, which lives only in sea water, West Indian and West African manatees are able to live in both salt water and freshwater. This means they can be found not only in the sea but in rivers as well. West African manatees have been seen more than 1000 miles (1600 km) from the sea in large African rivers.

The Amazonian manatee, on the other hand, cannot survive in salt water at all. It is found only in the lakes and rivers of the Amazon region of South America. The Amazonian manatee is also known as the South American manatee and is the smallest of the three species. It grows to only about 9 feet (2.7 m) long and weighs about 1000

Areas where manatees can be found

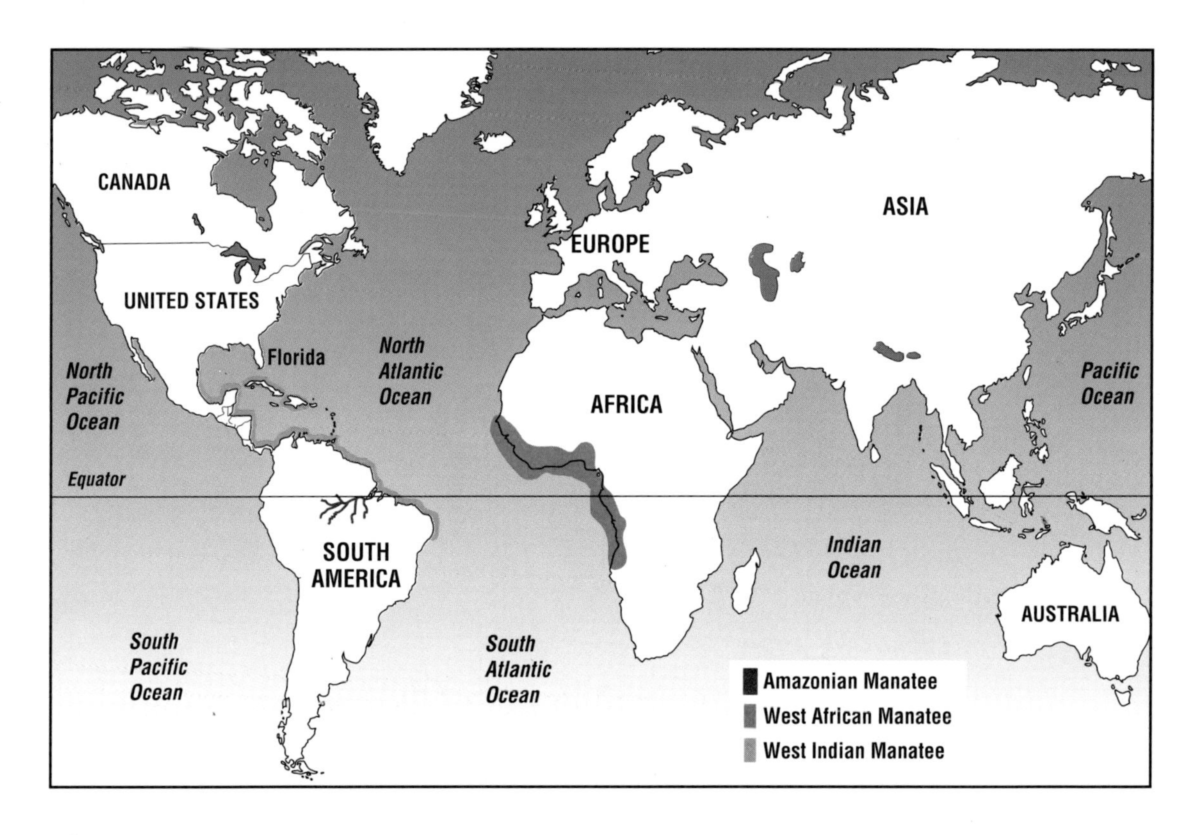

pounds (450 kg). The Amazonian manatee is a very dark blue-black with a pink belly patch and it has no nails.

Like dugongs, manatees are peaceful vegetarians that eat seagrasses and other water plants. The plants that manatees feed on are tough and difficult to **digest**, so their stomachs are full of **bacteria**. These tiny lifeforms help break down the tough vegetation. They may also protect the manatees from poisonous chemicals in the plants.

When they are feeding or steering in the water, manatees often use their flippers just as if they were arms. They also sometimes use their flippers for "walking" along the sea-floor or riverbed in shallow water. The West African and West Indian manatees do most of their feeding on the bed

Looking a bit like a pet dog, an Amazonian manatee peers at the camera. Amazonians seem to spend about eight hours a day eating and the remainder resting or swimming around.

of the lake, river, or sea, whereas the Amazonian manatee prefers to feed at the surface of the water on floating vegetation. Occasionally it will also feed on land plants that overhang the water from the riverbanks and can eat more than 30 pounds (14 kg) of food a day.

The Amazonian manatee must eat a lot during the wet season because during the dry season the water level can fall by 20 feet (6 m). This makes water plants very rare and difficult to find. By feeding heavily when there is plenty, the manatee builds up large amounts of blubber under its skin, which give it enough energy to survive up to six months without food.

Amazonian manatees browse on floating plants. These animals eat so much that people in the Amazon are using them to help keep canals and dams free of weeds.

West Indian manatees gather in the Crystal River, Florida. Each year, large numbers of manatees move into the river when it becomes too cold in the Gulf of Mexico.

Manatees are not particularly sociable animals and do not seem to live in permanent groups. However, they do seem to recognize other manatees and will greet one another by touching, often by mouth-to-mouth "kissing." Sometimes hundreds of manatees can be found feeding or resting together. Scientists do not believe these are organized groups. The manatees simply gather where there is plenty of food or where the water is warm.

There is one time when male manatees do get together, though, and that is when there are females around that are ready to mate. Instead of claiming territories and waiting for females to pass through, as male dugongs do, male manatees form groups and go in search of females. When

A West Indian manatee calf drinks milk from a teat just behind its mother's flipper.

they find one, the males quickly surround her. They push and shove one another violently, each male trying to show how fit and strong he is so that the female will mate with him. There may be 3 to17 males in a mating herd, and the female may choose to mate with several of them.

About a year later, the female gives birth to a single calf, which is similar in size to a baby dugong. Like dugong calves, young manatees feed only on their mother's milk to start with, but within a few weeks they begin to eat plant food. Seagrasses contain an extremely hard material called silica (SIL-i-ca) that grinds teeth down. Manatees often take in sand with their food, too, which also wears their teeth out. Because of this, the animals have to replace their teeth regularly, which they do in a special way.

Manatees are born with four rows of teeth, two in the upper jaw and two in the lower. As soon as the young manatee begins to eat plant food, these rows start to move forward in the jaw. They move very slowly – at a rate of a fraction of an inch a month. Eventually, the teeth reach the front of the mouth and drop out, leaving behind a whole new set! This process is repeated over and over again throughout the manatee's life. Like dugongs, manatees live a long time, and some may reach 40 years of age.

Like dugong calves, young manatees stay with their mothers until they are about two years old. During that time, calves and mothers have a close relationship.

Sirenians & Man

Manatees and dugongs have always been hunted in small numbers by local people for their meat. In South America, people like to eat the meat of the Amazonian manatee cooked in its own blubber. And in the Caribbean, the West Indian manatee is eaten at Christmas. However, sirenians have also been hunted commercially on a huge scale not only for their meat but also for the oil from their blubber. This began in the seventeenth century, and more and more

Besides being killed for food and oil, manatees were hunted for their hides, which are about 2 inches (5 cm) thick.

manatees and dugongs were killed. The slaughter reached a peak during the first half of the twentieth century. Between 1935 and 1954, about 200,000 manatees and dugongs were caught and killed.

Today, large-scale hunting has stopped, but manatees and dugongs continue to suffer. **Pollution** is a big problem for sirenians because the plants they feed on will not grow in dirty water, so the animals starve to death. In many countries, people have built cities all along the coast. These cities not only cause pollution but people have also had to destroy manatee and dugong habitat in order to build them.

Sirenians are also killed by boats. In Florida, about 50 West Indian manatees are killed in boating accidents each

This West Indian manatee has had its tail badly injured by a boat propeller. Manatees are curious, but they are also slow and cannot get out of the way of fast boats.

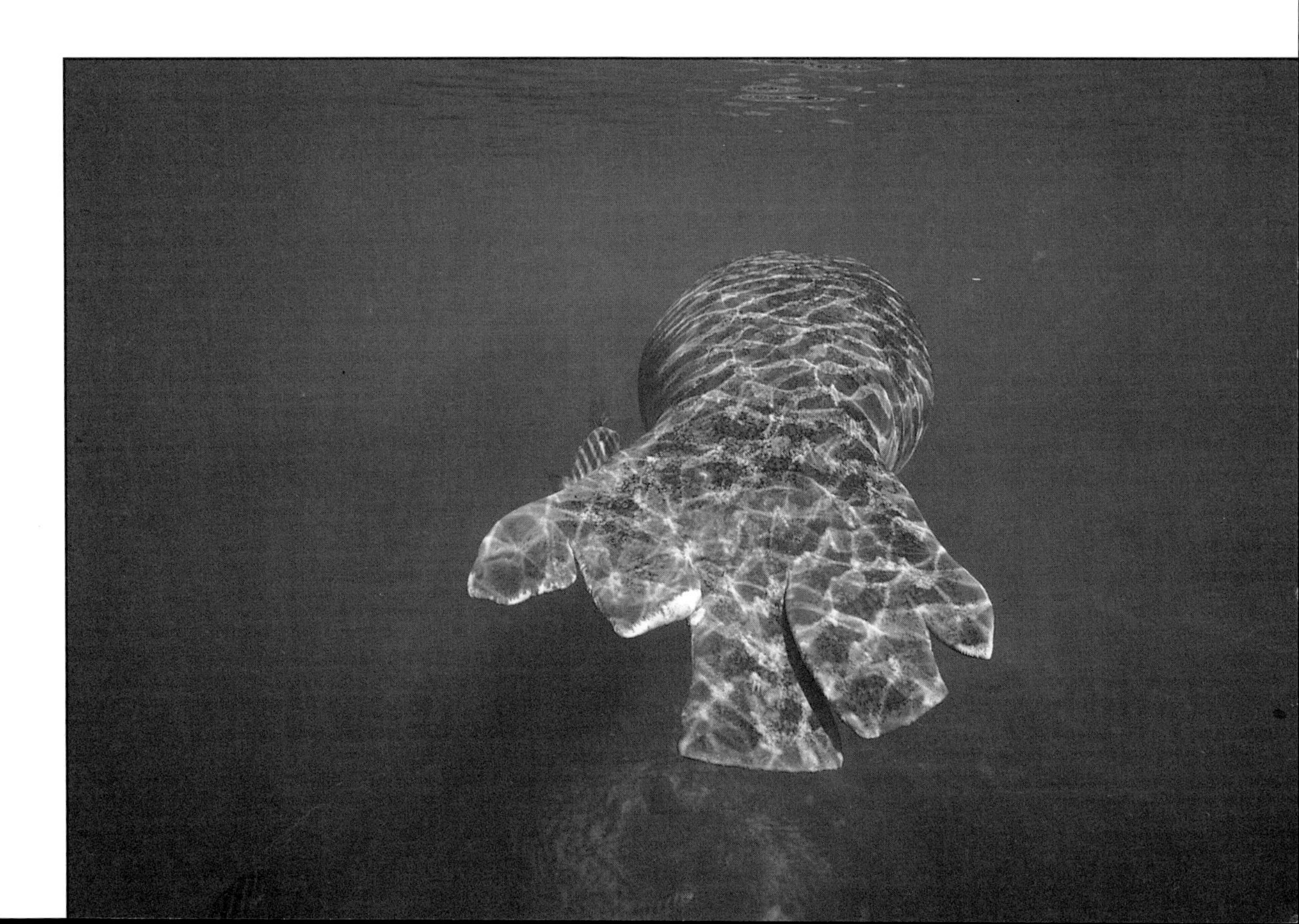

year. Besides those that are killed, many manatees are horribly injured by the blades of boat propellers.

Other dangers faced by sirenians include becoming trapped in fishing nets and nets used for protecting beaches from sharks. In just 20 years, between the early 1970s and the early 1990s, 522 dugongs were accidentally caught and drowned in shark nets off eastern Australia.

In spite of all this, sirenians have fascinated people for centuries. Some people believe that the mermaids that sailors thought they saw were really manatees and dugongs. And in Indonesia, a family even kept a pet dugong.

One day, Sirejudin, a fisherman, went to check his net and found a dugong trapped in it. Sadly, it was dead, but he

Sirejudin collects Maya's tears with a spoon. She is not crying because she is sad – a dugong's eyes water to keep them from getting too dry.

noticed its calf swimming round and round the boat. He took her home and named her Maya. During the day, the young dugong fed in the water, tied to a rope so that she did not swim off and get lost. At night, the family carried her into the house. Dugongs can survive out of water for a few hours, and in the house Maya was safe from hunters.

The family took great care of Maya. In return, each evening they collected her tears, which Indonesians believe bring good fortune. Once she was fully grown, though, the family had to let Maya go. Today, with luck, she is alive and well in the wild.

Once Maya had eaten all the water plants in one area, the family put her in a boat and moved her to new feeding grounds.

Saving Manatees & Dugongs

Manatees often swim very close to the surface. In Florida, signs warn boaters that there are manatees in the water and ask them to be careful.

Sadly, as a result of overhunting, loss of habitat, and fishing and boating accidents, scientists think that all species of sirenians are at risk of becoming extinct. However, because they live in water, it is extremely difficult to tell exactly how many manatees and dugongs survive in the wild. It is often especially hard to see those that live in the muddy waters of rivers and lakes. Even so, scientists estimate that there may be as few as 2600 West Indian manatees left in Florida waters and perhaps only 30,000 dugongs left in the whole world. They also know that manatees and dugongs have disappeared completely from some areas where they used to live. As we have seen, the dugong is probably extinct in the waters around Madagascar. It may possibly have vanished from those around Sri Lanka, too.

Many people are trying to help these gentle and harmless creatures, however. The **conservation** group known as the World Conservation Union has an organization called the Sirenia Specialist Group, which looks after manatees and dugongs. In 1983, this organization put forward two programs to help save sirenians. First, it recommended that **reserves** should be set up in areas where manatees and dugongs regularly feed and give birth. Second, it suggested that people should be taught about the problems faced by

In 1991, conservation workers began traveling around the Amazon, showing videos about Amazonian manatees to persuade the local people not to hunt them.

manatees and dugongs so that they would stop hunting them and harming them with their fishing nets and boats.

Today manatees and dugongs are protected in many reserves around the world. These include the Mamirauá Ecological Reserve and a number of other reserves in the Brazilian part of the Amazon rainforest, the Great Barrier Reef Marine National Park in northeastern Australia, and the Paradise Islands National Park in Mozambique, east Africa. In Brazil, the state environment agency known as Ibama is helping to clean up polluted waters and restore manatee habitat.

In the United States, the West Indian manatee is protected under federal law, and there is a Manatee Recovery Plan to help save it.

Nevertheless, sirenians still face many dangers. Some are natural. In 1996, for example, many West Indian manatees in Florida waters died from a mysterious disease. However, the main problems for manatees and dugongs are human disturbance and the changes that are made to their watery world. If sirenians are to survive, more manatee and dugong habitat needs to be preserved and the animals given the peace and quiet they need.

Two manatees toddle along a riverbed on their flippers. Sirenians need our help now and in future if they are to survive.

Useful Addresses

For more information about manatees and dugongs and how you can help protect them, contact these organizations:

Center for Marine Conservation
1725 De Sales Street NW
Suite 500
Washington, D.C. 20036

The Cousteau Society
870 Greenbrier Circle
Suite 402
Chesapeake, Virginia 23320

Save the Manatee Club
500 N. Maitland Avenue
Maitland, Florida 32751

U.S. Fish and Wildlife Service
Endangered Species and Habitat Conservation
400 Arlington Square
18th and C Streets NW
Washington, D.C. 20240

World Wildlife Fund
1250 24th Street NW
Washington, D.C. 20037

Further Reading

The Atlas of Endangered Animals Steve Pollock (New York: Facts On File, 1993)

Endangered Animals Lynn M. Stone (Chicago: Childrens Press, 1984)

Endangered Wildlife of the World (New York: Marshall Cavendish Corporation, 1993)

Manatees The Cousteau Society (New York: Little Simon, 1993)

Manatees Emilie U. Lepthien (Chicago: Childrens Press, 1991)

Manatees on Location Kathy Darling (New York: Lothrop, Lee & Shepard, 1991)

The Vanishing Manatee Margaret Goff Clark (New York: Cobblehill Books, 1990)

Wildlife of the World (New York: Marshall Cavendish Corporation, 1994)

Glossary

Adapt: To change in order to survive in new conditions.

Bacteria (Back-TEAR-ee-a): Tiny lifeforms that live in the stomachs of animals, as well as in many other places.

Conservation (Kon-ser-VAY-shun): Protecting and preserving the Earth's natural resources, such as animals, plants, and soil.

Digest: To turn food in the stomach into substances that an animal's body needs to survive.

Extinct (Ex-TINKT): No longer living anywhere in the world.

Habitat: The place where an animal lives. For example, the dugong's habitat is warm seas.

Herbivore (HER-biv-or): A kind of animal that eats plants rather than animals.

Mammal: A kind of animal that is warm-blooded and has a backbone. Most mammals are covered with fur or have hair. Females have glands that produce milk to feed their young.

Mate: When a male and female get together to produce young.

Pollution (Puh-LOO-shun): Materials, such as garbage, fumes, and chemicals, that damage the environment.

Range: The area in the world in which a particular species of animal can be found.

Reserve: Land that has been set aside where plants and animals can live without being harmed.

Species: A kind of animal or plant. For example, the dugong is a species of sirenian.

Suckle: When a baby animal drinks milk from its mother's teats.

Territory: The piece of land or water in which an animal lives. Male dugongs defend their territories against other males.

Tropical: Having to do with or found in the tropics, the warm region of the Earth near the Equator.

Vegetarian: An animal that eats only plants and plant parts, such as seeds, nuts, and fruit.

Index